The **ESSENTIALS**® of

Calculus II

D0710169

Staff of Research & Education Association

Research & Education Association
Visit our website at
www.rea.com

Research & Education Association
61 Ethel Road West
Piscataway, New Jersey 08854
E-mail: info@rea.com

THE ESSENTIALS®
OF CALCULUS II

Year 2006 Printing

Printed in the United States of America

Library of Congress Control Number 99-76781

International Standard Book Number 0-87891-578-8

What REA's Essentials®
Will Do for You

This book is part of REA's celebrated *Essentials*® series of review and study guides, relied on by tens of thousands of students over the years for being complete yet concise.

Here you'll find a summary of the very material you're most likely to need for exams, not to mention homework—eliminating the need to read and review many pages of textbook and class notes.

This slim volume condenses the vast amount of detail characteristic of the subject matter and summarizes the **essentials** of the field. The book provides quick access to the important facts, principles, theorems, concepts, and equations of the field.

It will save you hours of study and preparation time.

This *Essentials*® book has been prepared by experts in the field and has been carefully reviewed to ensure its accuracy and maximum usefulness. We believe you'll find it a valuable, handy addition to your library.

Larry B. Kling
Chief Editor

CONTENTS

This book covers the usual course outline of Calculus II. Earlier/basic topics are covered in *"THE ESSENTIALS OF CALCULUS I"*. More advanced topics will be found in *"THE ESSENTIALS OF CALCULUS III"*.

Chapter No.   **Page No.**

1   TECHNIQUES OF INTEGRATION   1

1.1   Table of Integrals   1
1.2   Integration by Parts   3
1.3   Partial Fractions   4
1.4   Trigonometric Substitution   5
1.4.1   General Rules for Trigonometric
               Substitutions   5
1.4.2   Summary of Trigonometric Substitutions   7
1.5   Quadratic Functions   8

2   APPLICATIONS OF THE INTEGRAL   10

2.1   Area   10
2.2   Volume of a Solid of Revolution   11
2.3   Work   14
2.4   Fluid Pressure   15
2.5   Area of Surface of Revolution   16
2.6   Arc Length   17

3   PARAMETRIC EQUATIONS   18

3.1   Parametric Equations   18
3.2   Derivatives of Parametric Equations   19
3.3   Arc Length   20

4   POLAR COORDINATES 22

4.1	Polar Coordinates	22
4.2	Graphs of Polar Equations	23
4.2.1	Rules of Symmetry	24
4.3	Polar Equation of Lines, Circles and Conics	27
4.3.1	Lines	27
4.3.2	Circles	27
4.3.3	Conics	28
4.4	Areas in Polar Coordinates	28

5   ANALYTIC GEOMETRY 31

5.1	Three-Dimensional Coordinate System	31
5.1.1	Three-Dimensional Distance Formula	32
5.1.2	The Formula of a Midpoint	32
5.2	Direction Cosine, Angle and Numbers	32
5.2.1	Directed Line	32
5.2.2	Direction Cosine	33
5.2.3	Direction Numbers	33
5.3	Equations of a Line and Plane in Space	34
5.3.1	Line	34
5.3.2	Plane	35

6   TWO AND THREE DIMENSIONAL VECTOR ANALYSIS 37

6.1	Two-Dimensional Vectors	37
6.1.1	Vector Properties	38
6.1.2	Additive and Multiplicative Properties of Vectors	39
6.1.3	Scalar (DOT) Product	40
6.2	Three-Dimensional Vectors	41
6.2.1	Vector Properties	41
6.2.2	Linear Dependence and Independence	42
6.3	Vector Multiplication	43
6.3.1	Scalar (DOT) Product	43
6.3.2	Vector (CROSS) Product	44
6.3.3	Product of Three Vectors	47

CHAPTER 1

TECHNIQUES OF INTEGRATION

1.1 TABLE OF INTEGRALS

1. $\int \alpha\, dx = \alpha x + C$.

2. $\int x^n dx = \frac{1}{n+1} x^{n+1} + C, \quad n \neq -1$.

3. $\int \frac{dx}{x} = \ln |x| + C$.

4. $\int e^x dx = e^x + C$.

5. $\int p^x dx = \frac{p^x}{\ln p} + C$.

6. $\int \ln x\, dx = x \ln x - x + C$.

7. $\int \cos x\, dx = \sin x + C$.

8. $\int \sin x\, dx = -\cos x + C$.

9. $\int \sec^2 x \, dx = \tan x + C.$

10. $\int \csc^2 x \, dx = -\cot x + C.$

11. $\int \sec x \tan x \, dx = \sec x + C.$

12. $\int \csc x \cot x \, dx = -\csc x + C.$

13. $\int \tan x \, dx = \ln |\sec x| + C.$

14. $\int \cot x \, dx = \ln |\sin x| + C.$

15. $\int \sec x \, dx = \ln |\sec x + \tan x| + C.$

16. $\int \csc x \, dx = \ln |\csc x - \cot x| + C.$

17. $\int \dfrac{dx}{\sqrt{1-x^2}} = \arcsin x + C.$

18. $\int \dfrac{dx}{1+x^2} = \arctan x + C.$

19. $\int \arcsin x \, dx = x \arcsin x + \sqrt{1-x^2} + C.$

20. $\int \arctan x \, dx = x \arctan x - \frac{1}{2} \ln (1+x^2) + C.$

21. $\int \dfrac{dx}{x^2-1} = \frac{1}{2} \ln \left| \dfrac{x-1}{x+1} \right| + C.$

2

1.2 INTEGRATION BY PARTS

Differential of a product is represented by the formula

$$d(uv) = udv + vdu$$

Integration of both sides of this equation gives

$$uv = \int udv + \int vdu \qquad (1)$$

or

$$\boxed{\int udv = uv - \int vdu} \qquad (2)$$

Equation (2) is the formula for integration by parts.

Example: Evaluate $\int x \ln x \, dx$

Let $\qquad u = \ln x \qquad dv = xdx$

$$du = 1/x \, dx \qquad v = 1/2 \, x^2$$

Thus,

$$\int x \ln x \, dx = (\tfrac{1}{2})x^2 \ln x - \int (\tfrac{1}{2})x^2 \cdot (\tfrac{1}{x})dx$$

$$= (\tfrac{1}{2})x^2 \ln x - \tfrac{1}{2} \int x \, dx$$

$$= (\tfrac{1}{2})x^2 \ln x - (\tfrac{1}{4})x^2 + c$$

Integration by parts may be used to evaluate definite integrals. The formula is:

$$\boxed{\int_a^b udv = \left[uv\right]_a^b - \int_a^b vdu}$$

3

1.3 PARTIAL FRACTIONS

To evaluate rational functions of the form $\int \frac{P(x)}{Q(x)}\,dx$, where P and Q are polynomials, we apply the following techniques:

1. Factor the denominator, $Q(x)$, into a product of linear and quadratic factors.
    Example: $Q(x) = x^3 + 2x^2 + x + 2$

    $$= x^3 + x + 2x^2 + 2 = x(x^2+1) + 2(x^2+1)$$

    $$= \underset{\substack{\text{quadratic}\\\text{factor}}}{(x^2+1)}\underset{\substack{\text{linear}\\\text{factor}}}{(x+2)}$$

2. Rewrite $\frac{P}{Q}$ as a sum of simpler rational functions, each of which can be integrated.

    If the degree of the numerator ($P(x)$) is larger than the degree of the denominator ($Q(x)$), we divide $P(x)$ by $Q(x)$ to obtain a quotient (polynomial of the form $\frac{P}{Q}$) plus a rational function (remainder divided by the divisor) in which the degree of the numerator is less than the degree of the denominator.

    The decomposition of a rational function into the sum of simpler expressions is known as the method of partial fractions. Four ways in which the denominator can be factored are as follows:

1. The denominator $Q(x)$ can be decomposed to give distinct linear factors of the form $\frac{A_1}{x-a_1} + \frac{A_2}{x-a_2} + \ldots + \frac{A_n}{x-a_n}$.

2. The denominator $Q(x)$ can be decomposed into linear factors of the form $\frac{A_1}{x-a} + \frac{A_2}{(x-a)^2} + \ldots + \frac{A_k}{(x-a)^k}$, where some of the linear factors are repeated.

4

Example: Decomposition of $Q(x) = (x-2)^3$ gives $\dfrac{A_1}{x-2} + \dfrac{A_2}{(x-2)^2} + \dfrac{A_3}{(x-2)^3}$.

3. $Q(x)$ can be factored to give linear and irreducible quadratic factors. Each unrepeated quadratic factor has the form $\dfrac{Ax+B}{x^2+bx+c}$.

4. $Q(x)$ can be factored to give linear and quadratic factors where some of the quadratic factors can be repeated.

   In this case the fraction can be expressed as follows:

$$\frac{A_1x + B_1}{(x^2+bx+c)} + \frac{A_2x + B_2}{(x^2+bx+c)^2} + \ldots + \frac{A_nx + B_n}{(x^2+bx+c)^n}$$

1.4 TRIGONOMETRIC SUBSTITUTION

If the integral contains expressions of the form $\sqrt{a^2-x^2}$, $\sqrt{a^2+x^2}$ or $\sqrt{x^2-a^2}$, where $a > 0$, it is possible to transform the integral into another form by means of trigonometric substitution.

1.4.1 GENERAL RULES FOR TRIGONOMETRIC SUBSTITUTIONS

1. Make appropriate substitutions.

2. Sketch a right triangle.

3. Label the sides of the triangle by using the substituted information.

4. The length of the third side is obtained by use of the Pythagorean Theorem.

5. Utilize sketch, in order to make further substitutions.

A. If the integral contains the expression of the form $\sqrt{a^2-x^2}$, make the substitution $x = a \sin\theta$.

$$\sqrt{a^2-x^2} = \sqrt{a^2-a^2\sin^2\theta} = \sqrt{a^2(1-\sin^2\theta)} = \sqrt{a^2\cos^2\theta}$$

$$= a \cos\theta$$

In trigonometric substitution the range of θ is restricted. For example, in the sine substitution the range of θ is $-\pi/2 \leq \theta \leq \pi/2$. The sketch of this substitution is shown in Fig. 1-1.

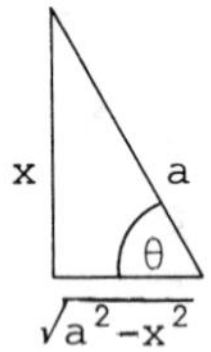

$$x = a \sin\theta, \text{ thus } \sin\theta = \frac{x}{a}$$

Fig. 1-1

B) If the integral contains the expression of the form $\sqrt{x^2-a^2}$, make the substitution $x = a \sec\theta$. The sketch is shown in Fig. 1-2

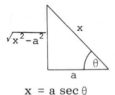

$$x = a \sec\theta$$

Fig. 1-2

C) If the integral contains the expression of the form $\sqrt{a^2+x^2}$, make the substitution $x = a \tan\theta$.

The sketch is shown in Figure 1-3.

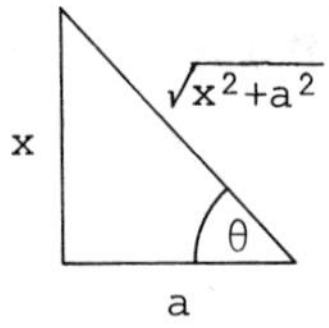

$$x = a \tan\theta$$

Fig. 1-3

Example: Evaluate $\displaystyle\int \frac{dx}{\sqrt{4 + x^2}}$

Solution: Let $x = 2\tan\theta$

$$dx = 2\sec^2\theta\, d\theta$$

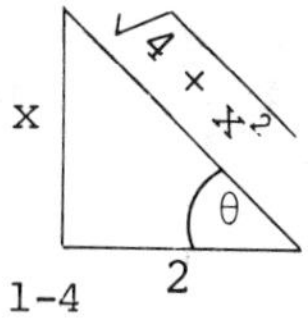

Fig. 1-4

Thus,

$$\int \frac{dx}{\sqrt{4 + x^2}} = \int \frac{2\sec^2\theta\, d\theta}{\sqrt{4 + (2\tan\theta)^2}}$$

$$= \int \frac{2\sec^2\theta\, d\theta}{\sqrt{4(1 + \tan^2\theta)}}$$

$$= \int \frac{2\sec^2\theta\, d\theta}{2\sqrt{\sec^2\theta}} = \int \sec\theta\, d\theta$$

$$= \ln|\sec\theta + \tan\theta| + c$$

To convert from θ back to x we use Fig. 1-4 to find:

$$\sec\theta = \frac{\sqrt{4 + x^2}}{2} \quad \text{and} \quad \tan\theta = \frac{x}{2}$$

Therefore, $\displaystyle\int \frac{dx}{\sqrt{4 + x^2}} = \ln\left|\frac{\sqrt{4 + x^2}}{2} + \frac{x}{2}\right| + c$

1.4.2 SUMMARY OF TRIGONOMETRIC SUBSTITUTIONS

Given expression	Trigonometric substitution
$\sqrt{x^2 - a^2}$	$x = a\sec\theta$
$\sqrt{x^2 + a^2}$	$x = a\tan\theta$
$\sqrt{a^2 - x^2}$	$x = a\sin\theta$

7

1.5 QUADRATIC FUNCTIONS

An integral containing the expression ax^2+bx+c can be simplified by completing the square and making the appropriate substitution.

Thus, $\quad ax^2 + bx + c = a(x^2 + \dfrac{b}{a}x) + c$

$$= a\left(x + \dfrac{b}{2a}\right)^2 + c - \dfrac{b^2}{4a}$$

Then substitute $y = x + \dfrac{b}{2a}$, which changes the expression into an integrable form.

Example: Evaluate $\displaystyle\int \dfrac{(2x-3)}{x^2+2x+2}\,dx$

Solution: We complete the square, obtaining

$$x^2 + 2x + 2 = (x^2 + 2x + 1) + 1$$

$$= (x+1)^2 + 1$$

Let $y = x+1$, $\quad dy = dx$

Thus,

$$\int \dfrac{(2x-3)\,dx}{x^2+2x+2} = \int \dfrac{2y-5}{y^2+1}\,dy = \int \dfrac{2y\,dy}{y^2+1} - 5\int \dfrac{dy}{y^2+1}$$

$$= \ln\,(y^2+1) - 5\,\arctan y + c$$

$$= \ln\,(x^2+2x+2) - 5\,\arctan(x+1) + c$$

The technique of completing the square may be used if the quadratic expression appears under the radical sign.

Example: Evaluate $\displaystyle\int \dfrac{1}{\sqrt{8+2x-x^2}}\,dx$

Solution: We complete the square, obtaining

$$8+2x-x^2 = 8-(x^2-2x) = 8+1-(x^2-2x+1)$$

$$= 9-(x-1)^2$$

Let $u = x-1$; $du = dx$

Thus,

$$\int \frac{1}{\sqrt{8+2x-x^2}} = \int \frac{1}{\sqrt{9-u^2}}\, du$$

$$= \sin^{-1}\frac{u}{3} + c$$

$$= \sin^{-1}\frac{x-1}{3} + c$$

$$= \arcsin \frac{x-1}{3} + c$$

CHAPTER 2

APPLICATIONS OF THE INTEGRAL

2.1 AREA

If f and g are two continuous functions on the closed interval [a,b], then the area of the region bounded by the graphs of these two functions and the ordinates x = a and x = b is

$$A = \int_a^b [f(x) - g(x)]dx.$$

where
$$f(x) \geq 0 \quad \text{and} \quad f(x) \geq g(x)$$

$$a \leq x \leq b$$

This formula applies whether the curves are above or below the x-axis.

The area below f(x) and above the x-axis is represented by $\int_a^b f(x)$. The area between g(x) and the x-axis is represented by $\int g(x)$.

Example: Find the area of the region bounded by the curves $y = x^2$ and $y = \sqrt{x}$.

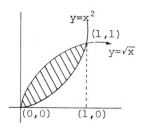

Fig. 2-1

$$\text{Area} = A = \int_0^1 (\sqrt{x} - x^2)\,dx$$

$$= \int_0^1 \sqrt{x}\,dx - \int_0^1 x^2\,dx$$

$$= \left[\frac{2}{3}\, x^{\frac{3}{2}} - \frac{1}{3}\, x^3 \right]_0^1$$

$$A = \left[\frac{2}{3} - \frac{1}{3} \right] = \frac{1}{3}$$

2.2 VOLUME OF A SOLID OF REVOLUTION

If a region is revolved about a line, a solid called a solid of revolution is formed. The solid is generated by the region. The axis of revolution is the line about which the revolution takes place.

There are several methods by which we may obtain the volume of a solid of revolution. We shall now discuss three such methods.

1. Disk Method

The volume of the solid generated by the revolution of a region about the x-axis is given by the formula

$$V = \pi \int_a^b [f(x)]^2 \, dx,$$

provided that f is a continuous, nonnegative function on the interval [a,b].

2. Shell Method

This method applies to cylindrical shells exemplified by

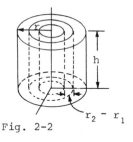

Fig. 2-2

The volume of a cylindrical shell is

$$V = \pi r_2^2 h - \pi r_1^2 h$$

$$= \pi (r_2 + r_1)(r_2 - r_1) h$$

$$= 2\pi \left(\frac{r_2 + r_1}{2} \right)(r_2 - r_1) h$$

where r_1 = inner radius

r_2 = outer radius

h = height.

Let $r = \frac{r_1 + r_2}{2}$ and $\Delta r = r_2 - r_1$, then

the volume of a shell becomes

$$V = 2\pi rh\Delta r$$

The thickness of the shell is represented by Δr and the average radius of the shell by r.

Thus,

$$V = 2\pi \int_a^b xf(x)\,dx$$

is the volume of a solid generated by revolving a region about the y-axis. This is illustrated by Fig. 2-3.

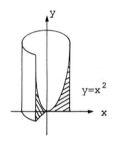

Fig. 2-3

3. Parallel Cross Sections

A cross section of a solid is a region formed by the intersection of a solid by a plane. This illustrated by Fig. 2-4.

Fig. 2-4

13

If x is a continuous function on the interval [a,b], then the volume of the cross sectional area A(x) is

$$V = \int_a^b A(x)dx.$$

2.3 WORK

Force is a physical property that is defined by Newton's law as the mass of an object multiplied by its acceleration:

$$F = ma.$$

where

F = force
m = mass
a = acceleration

An object that is subjected to a constant force which moves it a distance d in the direction of the force is said to have done work.

$$W = Fd$$

Variable forces are forces that are not constant.

Definition:

Let $f(x)$ represent the force at the point x along the x-axis, where f is continuous on the interval [a,b]. The work done in moving the object from a to b is

$$W = \int_a^b F(x)dx.$$

2.3.1 HOOKE'S LAW

The force $f(x)$ required to stretch a spring x units beyond its natural length is given by $f(x) = kx$, where k is a constant called the spring constant.

The same formula is used to find the work done in compressing a spring x units from its natural length.

2.4 FLUID PRESSURE

The pressure that a liquid exerts on a plate located at a depth h in a container is

$$P = \rho h$$

where ρ = density of the liquid (weight per unit volume)

         h = depth of the liquid.

From this, we may assume that the pressure of the liquid is dependent on the depth, but is independent of the size of the container.

Theorems:

1. The pressure is the same in all directions, at any point x, in a liquid.

2. The total force of a plate which is divided into several parts, is the sum of the forces on each of the parts.

The total force is defined as

$$F = \int_a^b \rho \, p(h) \, dh$$

where ρ is density of the liquid;
         p(h) is the pressure as a function of depth(h),
         a and b are the limits of the region in which
         the pressure is exerted.

2.5 AREA OF SURFACE OF REVOLUTION

A surface of revolution is generated when a plane is revolved about a line.

If f' and g' are two continuous functions on the interval [a,b] where g(t) = 0, x = f(t) and y = g(t) then, the surface area of a plane revolved about the x-axis is given by the formula

$$S = \int_a^b 2\pi \; g(t) \; \sqrt{[f'(t)]^2 + [g'(t)]^2}\,dt$$

Since x = f(t) and y = g(t),

$$S = \int_a^b 2\pi y \sqrt{\left(\frac{dx}{dt}\right)^2 + \left(\frac{dy}{dt}\right)^2}\,dt$$

If the plane is revolved about the y-axis, then the surface area is

$$S = \int_a^b 2\pi x \sqrt{\left(\frac{dx}{dt}\right)^2 + \left(\frac{dy}{dt}\right)^2}\,dt$$

These formulas can be simplified to give the following:

$$S = 2\pi y \int_a^b ds$$

for revolution about the x-axis, and

$$S = 2\pi x \int_a^b ds$$

for revolution about the y-axis.

In the above equations, ds is given as

$ds = \sqrt{1+f'(x)^2} \ dx$.

2.6 ARC LENGTH

If $f'(x)$ represents the derivative of a function f, and if $f'(x)$ is continuous, then the function is said to be smooth.

We shall now define the arc length of a smooth curve on the interval [a,b].

Definition:

The arc length of the graph of f from A(a,f(a)) to B(b,f(b)) is given by the formula

$$L = \int_a^b \sqrt{1+[f'(x)]^2} \ dx$$

if the function is smooth on the interval [a,b].

If g is a continuous function on the closed interval [c,d] and it is defined by $x = g(y)$, then the formula for the arc length is

$$L = \int_c^d \sqrt{1+[g'(y)]^2} \ dy$$

In this case, y is regarded as the independent variable.

CHAPTER 3

PARAMETRIC EQUATIONS

3.1 PARAMETRIC EQUATIONS

A parameter is a quantity whose value determines the value of other quantities. Given the equation $y = t^2+t+3$ and $x = t^2+2t$, where the value of t determines the value of x and y; t is considered a parameter.

Let x and y represent two differentiable functions on the interval [a,b] which are continuous at the endpoints and $(x(t),y(t))$ represents a point in the plane. If t is in the interval [a,b], then the curve traced out by the point $(x(t),y(t))$ is given parametrically as

$$x = x(t) \quad \text{and} \quad y = y(t)$$

To sketch the graph of equations given parametrically, we utilize the following steps:

1. Solve the first parametric equation in terms of t.

2. Substitute t into the second equation and simplify it.

3. Tabulate coordinates of the points on the curve.

4. Plot the point in the domain.

5. Sketch graph.

Example: Sketch the graph of the equations given parametrically as

$$x = t^2 + 4 \qquad\qquad (1)$$

$$y = t^2. \qquad\qquad (2)$$

From (1) we have $\sqrt{x-4} = t$. Hence

$$y = (\sqrt{x-4})^2 = x-4 \quad \text{i.e.}$$

$$x - y = 4$$

t	x	y
1	5	1
2	8	4
3	13	9
.		
.		
.		
-1	5	1
-2	8	4
-3	13	9

Here $t \in [-\infty, \infty]$.

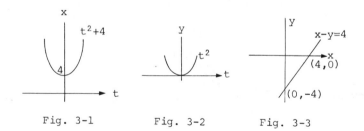

Fig. 3-1   Fig. 3-2   Fig. 3-3

3.2 DERIVATIVES OF PARAMETRIC EQUATIONS

Parametric equations can be used to obtain the slope of a curve (its derivative) by the computation of $\frac{dy}{dx}$, where x and y are given parametrically.

If $x = f(t)$ and $y = g(t)$, then

$$\frac{dy}{dx} = \frac{dy}{dt} \cdot \frac{dt}{dx} = \frac{dy/dt}{dx/dt}$$

is obtained by use of the Chain Rule. The derivative is given in terms of t.

We also use the Chain Rule to obtain the second derivative which is written as:

$$\frac{d^2 y}{dx^2} = \frac{d}{dx}\left(\frac{dy}{dx}\right) = \frac{d}{dt}\left(\frac{dy}{dx}\right)\frac{dt}{dx} = \frac{d}{dt}\left(\frac{dy}{dx}\right) \div \frac{dx}{dt}$$

The formula for the third derivative is,

$$\frac{d^3 y}{dx^3} = \frac{d}{dx}\left(\frac{d^2 y}{dx^2}\right) = \frac{d}{dt}\left(\frac{d^2 y}{dx^2}\right) \cdot \frac{dt}{dx} = \frac{\frac{d}{dt}\left(\frac{d^2 y}{dx^2}\right)}{dx/dt}$$

3.3 ARC LENGTH

If a graph is given by the parametric equation

$$x = f(t), \quad y = g(t) \quad a \le t \le b,$$

then f and g are continuous on the interval [a,b] and the parameters t_0 and t_1 have two different values.

The formula for arc length is

$$S = \int_a^b \sqrt{[x'(t)]^2 + [y'(t)]^2} \, dt$$

The differential of the arc length(s) is written as:

$$S'(t) = \frac{ds}{dt} = \sqrt{\left(\frac{dx}{dt}\right)^2 + \left(\frac{dy}{dt}\right)^2}$$

If the arc is in the form $y = f(x)$ or $x = g(y)$, then the length of the arc is

$$\frac{ds}{dx} = \sqrt{1 + \left(\frac{dy}{dx}\right)^2} \quad \text{or} \quad \frac{ds}{dy} = \sqrt{1 + \left(\frac{dx}{dy}\right)^2}$$

CHAPTER 4

POLAR COORDINATES

4.1 POLAR COORDINATES

Polar coordinates is a method of representing points in a plane by the use of ordered pairs.

The polar coordinate system consists of an origin (pole), a polar axis and a ray of specific angle.

The polar axis is a line that originates at the origin and extends indefinitely in any given direction.

The position of any point in the plane is determined by its distance from the origin and by the angle that the line makes with the polar axis.

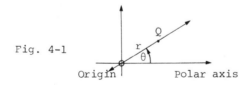

Fig. 4-1

The coordinates of the polar coordinate system are (r, θ).

The angle (θ) is positive if it is generated by a counterclockwise rotation of the polar axis, and is negative if it is generated by a clockwise rotation.

A point in the polar coordinate system can be represented by many $[r, \theta]$ pairs, thus,

1. if r = 0, then the resulting point has coordinates $[0, \theta]$, for all values of θ.

2. there is no difference between angles that differ by an integral multiple of 2π. Consequently,

$$[r, \theta] = [r, \theta + 2n\pi]$$ for all integers n.

3. to change the sign of the first coordinate, add π to the second coordinate:

$$[r, \theta + \pi] = [-r, \theta]$$

The relationship between polar coordinates (r, θ) and Cartesian coordinates (x, y) is given by the following equations:

1. if r and θ are given, then

$$x = r \cos\theta \quad y = r \sin\theta$$

2. if given x and y

$$\tan\theta = y/x, \quad r = \sqrt{x^2 + y^2} \quad \text{or} \quad r^2 = x^2 + y^2$$

$$\sin\theta = \frac{y}{\sqrt{x^2 + y^2}}, \quad \cos\theta = \frac{x}{\sqrt{x^2 + y^2}}$$

4.2 GRAPHS OF POLAR EQUATIONS

The graph of an equation in polar coordinates is a set of all points, each of which has at least one pair of polar coordinates (r, θ), which satisfies the given equation.

To plot a graph:

1. Construct a table of values of θ and r.

2. Plot these points.

3. Sketch the curve.

4.2.1 RULES OF SYMMETRY

1. The graph is symmetric with respect to the x-axis, if the substitution of $(r, -\theta)$, for (r, θ) gives the same equation.
2. The graph is symmetric with respect to the y-axis, if the substitution of $(r, \pi - \theta)$ for (r, θ) gives the same equation.

3. The graph is symmetric with respect to the origin, if the substitution of $(-r, \theta)$ or $(r, \theta + \pi)$ for (r, θ) gives the same equation.

The following graphs illustrate equations in polar coordinates.

1. If $r = a\cos\theta$, then (r, θ) describes a circle. The graph is a circle with diameter a. It is symmetric with respect to the x-axis.

Fig. 4-2

2. $r = a\sin\theta$
    The graph is a circle symmetric with respect to the y-axis.

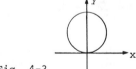

Fig. 4-3

3. $r = a$
    The graph is a circle symmetric with respect to the origin.

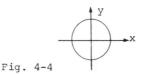

Fig. 4-4

4. $r = a \pm b \cos \theta$, $r = a \pm b \sin \theta$
    These graphs are called limancons.

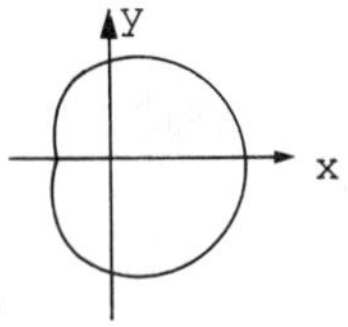

Fig. 4-5

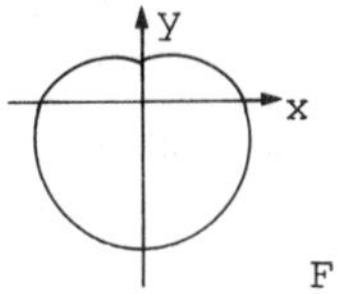

Fig. 4-6

$r = a + b \cos \theta$ where $a > b$

This graph is symmetric with respect to the x-axis.

$r = a - b \sin \theta$ where $a > b$

This graph is symmetric with respect to the y-axis.

5. The special case occurs when $a = b$. The graph is then called a Cardioid.

$r = a \pm b \cos \theta$ $r = a \pm b \sin \theta$.

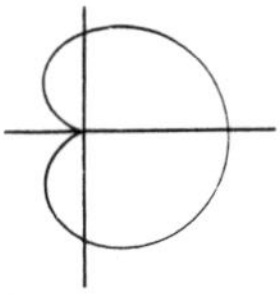

Fig. 4-7

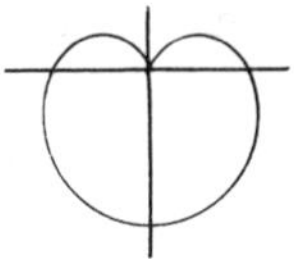

$r = a + b \cos \theta$
where $a = b$

Fig. 4-8

$r = a - b \sin \theta$
where $a = b$.

6. $r = a \cos n\theta$ $r = a \sin n\theta$
    These graphs are called rose or petal curves.

    The number of petals is equal to n if n is an odd integer and is equal to 2n if n is an even integer. If n is equal to one (n=1), there is one petal and it is circular.

Example: The graph of $r = 2 \cos 2\theta$ is illustrated below:

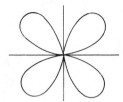

Fig. 4-9

7. $r^2 = a^2 \cos 2\theta \quad r^2 = a^2 \sin 2\theta$
   These are called lemniscates. The graphs are illustrated below.

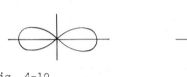

Fig. 4-10

8. The spiral of Archimedes has an equation of the form $r = k\theta$. The graph is illustrated below.

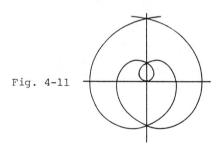

Fig. 4-11

9. The Logarithmic spiral has an equation of the form

$$\log r = \log a + k\theta$$

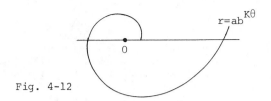

$r=ab^{K\theta}$

Fig. 4-12

4.3 POLAR EQUATION OF LINES, CIRCLES AND CONICS

4.3.1 LINES

The equation of a line in Cartesian coordinates is

$\alpha x + \beta y + \gamma = 0$, where α, β and γ are all constants.

To transform this equation to polar coordinates we substitute $x = r \cos\theta$ and $y = r \sin\theta$. The resulting equation is

$$-\frac{\gamma}{r} = \alpha \cos\theta + \beta\sin\theta$$

4.3.2 CIRCLES

The equation of a circle in polar coordinates is

$$r^2 - 2cr \cos(\theta - \alpha) + c^2 = a^2,$$

with radius a and center at (c, α)

4.3.3 CONICS

If the graph of a point moves so that the ratio of its distance from a fixed point to its distance from a fixed line remains constant, then the following theorems are true:

1. If the ratio is equal to one, then the curve is a parabola.

2. If the ratio is between 0 and one, then the curve is an ellipse.

3. If the ratio is greater than one, then the curve is a hyperbola.

Thus, the equation of a conic in polar coordinates is given by the formula

$$r = \frac{be}{1 \pm e\cos\theta} \; , \quad r = \frac{be}{1 \pm e\sin\theta}$$

The conic is a parabola if $e = 1$, an ellipse if $0 < e < 1$ or a hyperbola if $e > 1$.

4.4 AREAS IN POLAR COORDINATES

The area (A) of a region bounded by the curve $r = f(\theta)$, and by the lines $\theta = a$ and $\theta = b$ is given by the formula:

$$A = \frac{1}{2} \int_a^b r^2 d\theta = \frac{1}{2} \int_a^b [f(\theta)]^2 \, d\theta$$

When finding the area,

1. Sketch graph of the polar equations given.

2. Shade region for which area is sought.

3. Determine limits.

4. Solve using the equation for area.

Example: Find the area outside the circle $r = 2a \cos \theta$ and inside the cardioid $r = a(1+\cos\theta)$.

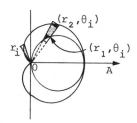

Fig. 4-13

Solution:

The required area is split evenly into two equal parts, above and below the x-axis. Therefore, to obtain the total area we can multiply the value for the area above the x-axis by 2. The area above the x-axis may be split into two parts and expressed as the sum of the area in the first quadrant (area of the cardioid minus the area of the circle, where θ goes from 0 to $\pi/2$) and the area of the second quadrant (the area of the cardioid alone where θ goes from $\pi/2$ to π).

Thus,

$$\text{Area} = A = \tfrac{1}{2} \int_a^b r^2 d\theta = \tfrac{1}{2} \int_a^b [f(\theta)]^2 \, d\theta$$

$$A_{\text{Total}} = 2 \int_0^{\frac{\pi}{2}} \tfrac{1}{2}[a(1+\cos\theta)]^2 \, d\theta - 2 \int_0^{\frac{\pi}{2}} \tfrac{1}{2}(2a\cos\theta)^2 \, d\theta$$

$$+ 2 \int_{\frac{\pi}{2}}^{\pi} \tfrac{1}{2}[a(1+\cos\theta)]^2 d\theta$$

29

$$A_{Total} = a^2 \int_0^{\frac{\pi}{2}} (1+2\cos\theta+\cos^2\theta)\,d\theta - a^2 \int_0^{\frac{\pi}{2}} 4\cos^2\theta\,d\theta$$

$$+ a^2 \int_{\frac{\pi}{2}}^{\pi} (1+2\cos\theta+\cos^2\theta)\,d\theta$$

$$A_{Total} = a^2 \int_0^{\frac{\pi}{2}} (1+2\cos\theta-3\cos^2\theta)\,d\theta$$

$$+ a^2 \int_{\frac{\pi}{2}}^{\pi} (1+2\cos\theta+\cos^2\theta)\,d\theta$$

Substitute $\cos^2\theta = \dfrac{1+\cos2\theta}{2}$ and integrate to obtain

$$A_{Total} = a^2 \left[2\sin\theta - \frac{\theta}{2} - \frac{3\sin2\theta}{4} \right]_0^{\frac{\pi}{2}}$$

$$+ a^2 \left[2\sin\theta + \frac{3\theta}{2} + \frac{\sin2\theta}{4} \right]_{\frac{\pi}{2}}^{\pi}$$

$$A_{Total} = a^2(2-\pi/4) + a^2\left(\frac{3\pi}{4} - 2 \right) = \frac{\pi a^2}{2}.$$

CHAPTER 5

ANALYTIC GEOMETRY

5.1 THREE-DIMENSIONAL COORDINATE SYSTEM

A set of ordered triples of real numbers is called the three-dimensional number space (R_3).

An ordered triple is composed of three numbers called coordinates.

Two ordered triples, (x_0, y_0, z_0) and $(x_1, y_1, z_1,)$, are equal if and only if $x_0 = x_1$, $y_0 = y_1$ and $z_0 = z_1$.

The coordinate axes are composed of three mutually perpendicular lines that intersect at the point $(0, 0, 0)$, called the origin. The axes form a right-hand system if the positive direction of the z- and y-axes lies in the plane of the paper, and if the positive direction of the x-axis is projected out of the plane of the paper. This is illustrated by Figure 5-1.

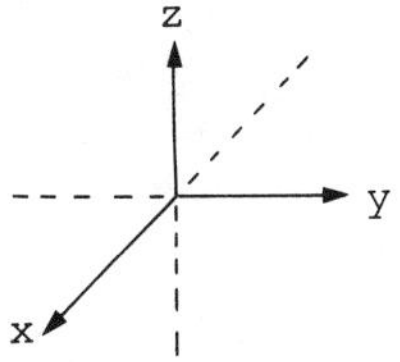

Fig. 5-1

If the y- and x-axes are interchanged, then the axes form a left-hand system.

A coordinate plane is a plane containing two of the coordinate axes. A Cartesian or rectangle coordinate system is a system in which each point in R_3 space has only one ordered number triple, and each ordered number triple designates only one point in R_3 space.

5.1.1 THREE-DIMENSIONAL DISTANCE FORMULA

The distance (d) between the two points $P_1(x_1,y_1,z_1)$ and $P_2(x_2,y_2,z_2)$ is given by the formula,

$$d = \sqrt{(x_2-x_1)^2 +(y_2-y_1)^2 +(z_2-z_1)^2} .$$

5.1.2 THE FORMULA OF A MIDPOINT

The midpoint of the line segment that connects two points is,

$$x_m = \frac{x_1+x_2}{2}, \quad y_m = \frac{y_1+y_2}{2}, \quad z_m = \frac{z_1+z_2}{2}$$

5.2 DIRECTION COSINE, ANGLE AND NUMBERS

5.2.1 DIRECTED LINE

A line which passes through the origin and points in the positive direction of the plane is called a directed line. A directed line is denoted L. An undirected line does not have an arrow on it, i.e., its direction is unsigned.

Fig. 5-2

5.2.2 DIRECTION COSINE

A directed line in space is determined by three angles called direction angles.

The angle formed between a directed line and the coordinate axes is called a direction angle.

If the direction angles of $\vec{L}$ are α, β and γ, then direction cosines are defined as $\cos \alpha$, $\cos \beta$ and $\cos \gamma$. Thus,

$$\boxed{\cos^2\alpha + \cos^2\beta + \cos^2\gamma = 1}$$

Lines in space which have the same direction cosines are said to be parallel. If there is a non-negative integer k such that

$$a_2 = ka_1, \quad b_2 = kb_1, \quad c_2 = kc_1$$

then the two sets of numbered triples (a_1,b_1,c_1) and (a_2,b_2,c_2) are proportional. This occurs if the coordinate of all the numbered triples are not all zeros.

5.2.3 DIRECTION NUMBERS

Integers which are proportional to the direction cosine of a line are called direction numbers.

The set of all direction numbers of a line form an infinite set.

The sets of direction numbers on a line passing through points $P_1(x_1,y_1,z_1)$ and $P_2(x_2,y_2,z_2)$, are

$$\boxed{x_2 - x_1, \quad y_2 - y_1 \quad \text{and} \quad z_2 - z_1}$$

The direction cosines of the line through the points $P_1(x_1,y_1,z_1)$ and $P_2(x_2,y_2,z_2)$ are

$$\cos\alpha = \frac{x_2-x_1}{d}, \quad \cos\beta = \frac{y_2-y_1}{d}, \quad \cos\gamma = \frac{z_2-z_1}{d}$$

where d represents the distance from P_1 to P_2.

If the set of direction numbers of line L_1 is proportional to the set of direction numbers of line L_2, then the two lines are parallel.

The angle between two directed lines $L_1(\cos\alpha_1, \cos\beta_1, \cos\gamma_1)$ and $L_2(\cos\alpha_2, \cos\beta_2, \cos\gamma_2)$ is given by the formula

$$\cos\theta = \cos\alpha_1\cos\alpha_2 + \cos\beta_1\cos\beta_2 + \cos\gamma_1\cos\gamma_2$$

where $\cos\alpha$, $\cos\beta$ and $\cos\gamma$ represent the direction cosines.

Two lines with direction numbers a_1, b_1, c_1 and a_2, b_2, c_2, respectively, are perpendicular if and only if

$$a_1a_2 + b_1b_2 + c_1c_2 = 0.$$

5.3 EQUATIONS OF A LINE AND PLANE IN SPACE

5.3.1 LINE

The equations of a line joining two points (x_0, y_0, z_0) and (x_1, y_1, z_1) are expressed parametrically as

$$
\begin{aligned}
x &= x_0 + (x_1 - x_0)t \\
y &= y_0 + (y_1 - y_0)t \\
z &= z_0 + (z_1 - z_0)t
\end{aligned}
$$

If a, b and c are the direction numbers of a line through

34

the point $P_0(x_0, y_0, z_0)$, then the parametric equation is

$$x = x_0 + at, \quad y = y_0 + bt, \quad z = z_0 + ct$$

Utilizing direction cosines, the equation can be written as

$$x = x_0 + t\cos\alpha; \quad y = y_0 + t\cos\beta; \quad z = z_0 + t\cos\gamma$$

The equation of a line may also be written symmetrically as

$$\frac{x - x_0}{x_1 - x_0} = \frac{y - y_0}{y_1 - y_0} = \frac{z - z_0}{z_1 - z_0}$$

A line, therefore, is represented as the intersection of two planes.

5.3.2 PLANE

In the three-dimensional system, the equation of a plane is represented by the equation,

$$Ax + By + Cz + D = 0.$$

A plane is determined by three points, all of which are not present on the same straight line.

Let P_0 represent a point and L represent a line, then there is only one plane that can pass through P_0 and is perpendicular to L. If A, B and C are direction numbers, then the equation of the plane through P_0 and perpendicular to L is

$$A(x-x_0) + B(y-y_0) + C(z-z_0) = 0$$

All parallel lines are perpendicular to the same plane since their direction numbers are proportional.

A set of direction numbers of a line which are perpendicular to a plane are called a set of altitude numbers of that plane. If the altitude numbers of two planes are proportional, then the planes are said to be parallel.

Two planes are perpendicular if and only if

$$A_1A_2 + B_1B_2 + C_1C_2 = 0$$

where A_1, B_1, C_1 and A_2, B_2, C_2 are the altitude numbers of the planes.

Let $A_1x + B_1y + C_1z = 0$ and $A_2x + B_2y + C_2z = 0$ represent two planes.

The angle between these two planes is given by the formula

$$\cos \theta = \frac{\left| A_1A_2 + B_1B_2 + C_1C_2 \right|}{\sqrt{A_1^2 + B_1^2 + C_1^2} \; \sqrt{A_2^2 + B_2^2 + C_2^2}}$$

The distance from the plane $Ax + By + Cz + D = 0$ to the point $P(x,y,z)$, is

$$d = \frac{\left| Ax + By + Cz + D \right|}{\sqrt{A^2 + B^2 + C^2}}$$

CHAPTER 6

TWO-AND THREE-DIMENSIONAL VECTOR ANALYSIS

6.1 TWO-DIMENSIONAL VECTORS

Scalar quantity - quantity that can be specified by a real number, thus has magnitude, but no direction.

A vector quantity is a quantity that has both magnitude and direction. Velocity is an example of a vector quantity.

A vector (AB) is denoted $\overrightarrow{AB}$, where B represents the head and A represents the tail. This illustrated in Fig. 6-1.

Fig. 6-1

The length of a line segment is the magnitude of a vector. If the magnitude and direction of two vectors are the same, then they are equal.

Vectors which can be translated from one position to another without any change in their magnitude or direction are called free vectors.

The unit vector is a vector with a length (magnitude) of one.

The zero vector has a magnitude of zero.

The unit vector, $\vec{i}$, is a vector with magnitude of one in the direction of the x-axis.

The unit vector $\vec{j}$ is a vector with magnitude of one in the direction of the y-axis.

6.1.1 VECTOR PROPERTIES

1. When two vectors are added together, the resultant force of the two vectors produce the same effect as the two combined forces. This is illustrated in Fig. 6-2.

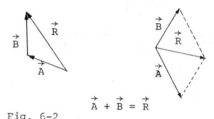

$$\vec{A} + \vec{B} = \vec{R}$$

Fig. 6-2

In these diagrams, the vector $\vec{R}$ is called the resultant vector.

If $\vec{A} = (a_1, a_2)$ and $\vec{B} = (b_1, b_2)$

Then,

$$\vec{A} + \vec{B} = \vec{R} = (a_1 + b_1, a_2 + b_2)$$

If c is a scalar constant then,

$$c\vec{A} = c(a_1, a_2) = (ca_1, ca_2)$$

6.1.2 ADDITIVE AND MULTIPLICATIVE PROPERTIES OF VECTORS

Let $\vec{S}$, $\vec{T}$, $\vec{U}$ represent vectors and let d and c represent constants. Thus,

1. $\vec{S} + \vec{T} = \vec{T} + \vec{S}$

2. $(\vec{S}+\vec{T})+\vec{U} = \vec{S}+(\vec{T}+\vec{U})$

3. $\vec{S} + 0 = \vec{S}$

4. $\vec{S} + (-\vec{S}) = 0$

5. $(c+d)\vec{S} = c\vec{S}+\vec{S}d$

6. $c(\vec{S}+\vec{U}) = c\vec{S} + c\vec{U}$

7. $c(\vec{S}\vec{T}) = (c\vec{S})\vec{T}$

8. $1 \cdot \vec{S} = \vec{S}$

9. $0 \cdot \vec{S} = 0$

10. $\vec{S} \cdot \vec{S} = |\vec{S}|^2$

11. $c(d\vec{S}) = (cd)\vec{S}$

12. $\vec{S} \cdot \vec{T} = \vec{T} \cdot \vec{S}$

The magnitude $|\vec{S}|$ of a vector $\vec{S} = a_1\vec{i} + a_2\vec{j}$ is

$$|\vec{S}| = \sqrt{a_1^2 + a_2^2}$$

The difference between vectors $\vec{A}$ and $\vec{B}$ is given by the formula

$$\vec{A} - \vec{B} = \vec{A} + (-\vec{B})$$

6.1.3 SCALAR (DOT) PRODUCT

Two vectors are parallel if one is a scalar multiple of the other and if neither is zero.

Definition:

If vector $\vec{A} = (a_1, a_2)$ and vector $\vec{B} = (b_1, b_2)$, then the scalar product of $\vec{A}$ and $\vec{B}$ is given by the formula,

$$\vec{A} \cdot \vec{B} = a_1 b_1 + a_2 b_2$$

Theorem:

If θ is the angle between the vectors $\vec{A} = a_1\vec{i} + b_1\vec{j}$ and $\vec{B} = a_2\vec{i} + b_2\vec{j}$ then,

$$\cos\theta = \frac{a_1 b_1 + a_2 b_2}{|\vec{A}||\vec{B}|}$$

Definition:

Let vector $\vec{A}$ be $a_1\vec{i} + a_1\vec{j}$ and vector $\vec{B} = b_2\vec{i} + b_2\vec{j}$. The projection of vector $\vec{A}$ on $\vec{B}$ ($\text{Proj}_{\vec{B}}\vec{A}$) is given by the quantity $|\vec{A}| \cos\theta$.

$$\text{Proj}_{\vec{B}}\vec{A} = |\vec{A}| \cos\theta = \frac{a_1 b_1 + a_1 b_1}{|\vec{B}|}$$

θ is the angle between the two vectors. If the angle θ is acute, then $|\vec{A}| \cos\theta$ is positive; if θ is obtuse, then $|\vec{A}| \cos\theta$ is negative.

The scalar product of two non-zero vectors, $\vec{A}$ and $\vec{B}$, is now redefined by the formula

$$\vec{A} \cdot \vec{B} = |\vec{A}||\vec{B}| \cos\theta = a_1 b_1 + a_2 b_2$$

6.2 THREE-DIMENSIONAL VECTORS

A vector in 3-space is denoted

$$\vec{v} = a\vec{i} + b\vec{j} + c\vec{k}$$

where i is the unit vector in the direction of the x-axis

j - is the unit vector in the direction of the y-axis

k - is the unit vector in the direction of the z-axis.

The magnitude of vector $\vec{v}$ is given by the formula

$$|v| = \sqrt{a^2+b^2+c^2}$$

The unit vector (u) in the direction of vector v is,

$$\vec{u} = \frac{1}{|\vec{v}|} \vec{v}$$

6.2.1 VECTOR PROPERTIES

1. Two vectors are proportional (parallel) if each vector is a scalar multiple of the other; there is a number c such that $\vec{u} = c\vec{v}$.

2. Two vectors $\vec{A}$ and $\vec{B}$ are orthogonal (perpendicular) if and only if $\vec{A} \cdot \vec{B} = 0$.

3. Two vectors, $\vec{AB}$ and $\vec{CD}$, are equal (have the same magnitude and direction) if and only if one of the following conditions are met.

   a. The two vectors are on the same directed line $\vec{L}$ and their directed lengths are equal; or

   b. The points A, B, C, D are the vertices of a parallelogram. This is illustrated in Fig. 6-3.

41

Fig. 6-3

6.2.2 LINEAR DEPENDENCE AND INDEPENDENCE

Two vectors are parallel if they are scalar multiples of each other; hence, there is a number c such that $\vec{u} = c\vec{v}$. Let $\vec{v}_1, \vec{v}_2, \ldots, \vec{v}_n$ represent a set of vectors and $c_1, c_2, \ldots, c_n$ represent numbers. The expression of the form

$$c_1\vec{v}_1 + c_2\vec{v}_2 + \ldots + c_n\vec{v}_n$$

is the linear combination of the vectors.

A set of vectors is linearly dependent if and only if there exists a set of constants such that

$$c_1\vec{v}_1 + c_2\vec{v}_2 + \ldots + c_n\vec{v}_n = 0$$

If these constants are all zero, then the set of vectors is said to be linearly independent.

Two proportional (parallel) vectors are linearly dependent; thus one member of the set can be expressed as a linear combination of the remaining members.

A set of vectors $\vec{r}$, $\vec{s}$, $\vec{t}$ are linearly independent if

$$\vec{r} = a_{11}\vec{i} + a_{12}\vec{j} + a_{13}\vec{k}$$

$$\vec{s} = a_{21}\vec{i} + a_{22}\vec{j} + a_{23}\vec{k}$$

$$\vec{t} = a_{31}\vec{i} + a_{32}\vec{j} + a_{33}\vec{k}$$

and the determinant $D = \begin{vmatrix} a_{11} & a_{12} & a_{13} \\ a_{21} & a_{22} & a_{23} \\ a_{31} & a_{32} & a_{33} \end{vmatrix}$

is not equal to zero.

The determinant is found by expanding the vector in the manner illustrated below:

$$\bar{u} = 2\bar{i} + \bar{j} - \bar{k}$$

$$\bar{v} = -\bar{i} + \bar{j} + 2\bar{k}$$

$$\bar{w} = 2\bar{i} - \bar{j} + 3\bar{k}$$

$$D = \begin{vmatrix} 2 & 1 & -1 \\ -1 & 1 & 2 \\ 2 & -1 & 3 \end{vmatrix} = 2\begin{vmatrix} 1 & 2 \\ -1 & 3 \end{vmatrix} - \begin{vmatrix} -1 & 2 \\ 2 & 3 \end{vmatrix} - \begin{vmatrix} -1 & 1 \\ 2 & -1 \end{vmatrix}$$

$$= 2(3+2)-(-3-4)-(1-2)$$

$$= 10 + 7 + 1 = 18$$

$$D = 18 \neq 0.$$

Thus the set of vectors is linearly independent.

6.3 VECTOR MULTIPLICATION

6.3.1 SCALAR (DOT) PRODUCT

Scalar (Dot) Product

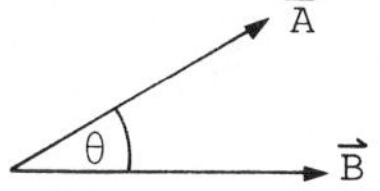

Fig. 6-4

Let $\vec{A} = a_1\bar{i} + a_2\bar{j} + a_3\bar{k}$ and $\vec{B} = b_1\bar{i} + b_2\bar{j} + b_3\bar{k}$. If θ is the angle between these two vectors, then

$$\cos \theta = \frac{a_1 b_1 + a_2 b_2 + a_3 b_3}{|\vec{A}|\,|\vec{B}|}$$

The scalar product of vectors A and $\vec{B}$ is

$$\vec{A} \cdot \vec{B} = |\vec{A}|\,|\vec{B}| \cos\theta = a_1 a_2 + b_1 b_2 + c_1 c_2$$

Definition:

Let θ represent the angle between vectors $\vec{A}$ and $\vec{B}$.

The component of $\vec{A}$ along $\vec{B}$ or the projection of $\vec{A}$ on $\vec{B}$ ($|\vec{A}|\cos\theta$) is given by the formula,

$$\text{Proj}_{\vec{B}}\vec{A} = |\vec{A}|\cos\theta = |\vec{A}|\,\frac{\vec{A}\cdot\vec{B}}{|A|\,|\vec{B}|} = \frac{\vec{A}\cdot\vec{B}}{|\vec{B}|}$$

Scalar products are used to calculate the work done by a constant force when its point of application moves along a line segment from $\vec{A}$ to $\vec{B}$. The work done is the product of the distance from $\vec{A}$ to $\vec{B}$ and the projection of the constant force F on vector $\overrightarrow{AB}$.

$$\text{Proj}_{W}F = \text{Proj}_{V}F = \frac{F \cdot v}{|v|}$$

The work done by F is given by the formula

$$W = F \cdot v$$

6.3.2 VECTOR (CROSS) PRODUCT

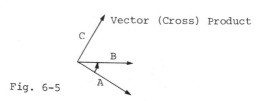

Fig. 6-5

44

The linearly independent vectors illustrated in Fig. 6-5 are said to form a right-handed triple.

The vectors in Fig. 6-6 form a left-handed triple.

Fig. 6-6

If two sets of ordered triples of vectors are both right-handed or left-handed, then they are said to be similarly oriented. If not, they are oriented in an opposite manner.

Theorem:

If the ordered triple $<A,B,C>$ is right-handed, then the ordered triples $<A,B,-C>$ and $<c_1A,c_2B,c_3C>$ are also right-handed provided that $c_1,c_2,c_3 > 0$.

Definition:

If A and B are vectors, then the vector product $A \times B$ is defined as follows:

1. if either A or B is 0, then

$$A \times B = 0$$

2. if A is parallel to B, then

$$A \times B = 0$$

3. otherwise

$$A \times B = C$$

where C has the following properties

45

a. it is orthogonal to both $\hat{A}$ and $\vec{B}$

b. it has magnitude $|\vec{C}| = |\hat{A}||\vec{B}|\sin\theta$, where θ is the angle betwen $\hat{A}$ and $\vec{B}$, and

c. it is directed so that $<\hat{A},\vec{B},\vec{C}>$ is a right-handed triple.

Theorem:

Let $\hat{A}$ and $\vec{B}$ represent vector, $<\vec{i},\vec{j},\vec{k},>$ represent a right-handed triple and t represent any number. Then

1. $\hat{A} \times \vec{B} = -(\vec{B} \times \hat{A})$

2. $(t\hat{A}) \times \vec{B} = t(\hat{A} \times \vec{B}) = \hat{A} \times (t\vec{B})$

3. $\vec{i} \times \vec{j} = -\vec{j} \times \vec{i} = \vec{k}$

4. $\vec{j} \times \vec{k} = -\vec{k} \times \vec{j} = \vec{i}$

5. $\vec{k} \times \vec{i} = -\vec{i} \times \vec{k} = \vec{j}$

6. $\vec{i} \times \vec{i} = \vec{j} \times \vec{j} = \vec{k} \times \vec{k} = 0$

Theorem:

If $\vec{A}$, $\vec{B}$ and $\vec{D}$ are vectors, then

1. $\vec{A} \times (\vec{B}+\vec{D}) = \vec{A} \times \vec{B} + \vec{A} \times \vec{D}$

2. $(\vec{A}+\vec{D}) \times \vec{B} = \vec{A} \times \vec{B} + \vec{A} \times \vec{D}$

Theorem:

If

$$\vec{A} = a_1\vec{i} + a_2\vec{j} + a_3\vec{k} \text{ and } \vec{B} = b_1\vec{i} + b_2\vec{j} + b_3\vec{k}$$

then

$$\vec{A} \times \vec{B} = \begin{vmatrix} \vec{i} & \vec{j} & \vec{k} \\ a_1 & a_2 & a_3 \\ b_1 & b_2 & b_3 \end{vmatrix}$$

$$\vec{A} \times \vec{B} = (a_2 b_3 - a_3 b_2)\vec{i} + (a_3 b_1 - a_1 b_3)\vec{j} + (a_1 b_2 - a_2 b_1)\vec{k}$$

Example: Find the cross product A × B, if

$$A = -2i + 4j + 5k \quad \text{and} \quad B = 4i + 5k$$

$$
\begin{array}{ccc}
20i & + \;20j & + \;0k \\
0i & + \;10j & - \;16k \\
\end{array}
$$

$$A \times B = 20i + 30j - 16k$$

When moving upwards, multiply by -1.

6.3.3 PRODUCT OF THREE VECTORS

Theorem:

If v_1, v_2, v_3 are vectors and the points PQRS are chosen so that

$$u(\overrightarrow{PQ}) = v_1 \quad u(\overrightarrow{PR}) = v_2 \quad u(\overrightarrow{PS}) = v_3,$$

then

1. $\left| (v_1 \times v_2) \cdot v_3 \right|$ is the volume of the parallelepiped with a vertex at P and adjacent vertices at Q, R and S. The volume is zero if and only if the four points P, Q, R and S lie in a plane.

2. if $<i,j,k>$ is a right-handed coordinate triple and if

$$v_1 = a_1 i + b_1 j + c_1 k,$$

$$v_2 = a_2 i + b_2 j + c_2 k,$$

$$v_3 = a_3 i + b_3 j + c_3 k,$$

then

$$(v_1 \times v_2) \cdot u_3 = \begin{vmatrix} a_1 & b_1 & c_1 \\ a_2 & b_2 & c_2 \\ a_3 & b_3 & c_3 \end{vmatrix}$$

3. $(v_1 \times v_2) \cdot v_3 = v_1 \cdot (v_2 \times v_3)$

Theorem:

Suppose A, B and D are any vectors, then

1. $(A \times B) \times D = (A - D)B - (B \cdot D)A$,

2. $A \times (B \times D) = (A \cdot D)B - (A \cdot B)D$.

HANDBOOK of
**MATHEMATICAL,
SCIENTIFIC, &
ENGINEERING**
FORMULAS, FUNCTIONS,
GRAPHS, TABLES,
& TRANSFORMS

RESEARCH & EDUCATION ASSOCIATION

A particularly useful reference for those in math, science, engineering and other technical fields. Includes the most-often used formulas, tables, transforms, functions, and graphs which are needed as tools in solving problems. The entire field of special functions is also covered. A large amount of scientific data which is often of interest to scientists and engineers has been included.

Available at your local bookstore or order directly from us by sending in coupon below.

Available at your local bookstore or order directly from us by sending in coupon below.

Available at your local bookstore or order directly from us by sending in coupon below.

Available at your local bookstore or order directly from us by sending in coupon below.

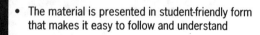

The High School Tutors®

The **HIGH SCHOOL TUTOR** series is based on the same principle as the more comprehensive **PROBLEM SOLVERS**, but is specifically designed to meet the needs of high school students. REA has revised all the books in this series to include expanded review sections and new material. This makes the books even more effective in helping students to cope with these difficult high school subjects.

If you would like more information about any of these books,
complete the coupon below and return it to us or go to your local bookstore.

"The ESSENTIALS" of Math & Science

Each book in the ESSENTIALS series offers all essential information of the field it covers. It summarizes what every textbook in the particular field must include, and is designed to help students in preparing for exams and doing homework. The ESSENTIALS are excellent supplements to any class text.

The ESSENTIALS are complete and concise with quick access to needed information. They serve as a handy reference source at all times. The ESSENTIALS are prepared with REA's customary concern for high professional quality and student needs.

Available in the following titles:

Advanced Calculus I & II
Algebra & Trigonometry I & II
Anatomy & Physiology
Anthropology
Astronomy
Automatic Control Systems /
    Robotics I & II
Biology I & II
Boolean Algebra
Calculus I, II, & III
Chemistry
Complex Variables I & II
Computer Science I & II
Data Structures I & II
Differential Equations I & II
Electric Circuits I & II
Electromagnetics I & II

Electronics I & II
Electronic Communications I & II
Fluid Mechanics /
    Dynamics I & II
Fourier Analysis
Geometry I & II
Group Theory I & II
Heat Transfer I & II
LaPlace Transforms
Linear Algebra
Math for Computer Applications
Math for Engineers I & II
Math Made Nice-n-Easy Series
Mechanics I, II, & III
Microbiology
Modern Algebra
Molecular Structures of Life

Numerical Analysis I & II
Organic Chemistry I & II
Physical Chemistry I & II
Physics I & II
Pre-Calculus
Probability
Psychology I & II
Real Variables
Set Theory
Sociology
Statistics I & II
Strength of Materials &
    Mechanics of Solids I & II
Thermodynamics I & II
Topology
Transport Phenomena I & II
Vector Analysis

*If you would like more information about any of these books,
complete the coupon below and return it to us or visit your local bookstore.*

RESEARCH & EDUCATION ASSOCIATION
61 Ethel Road W. • Piscataway, New Jersey 08854
Phone: (732) 819-8880   **website: www.rea.com**

Please send me more information about your Math & Science Essentials books

Name _____

Address _____

City _____ State _____ Zip _____

REA's **Problem Solvers**

The "PROBLEM SOLVERS" are comprehensive supplemental text-books designed to save time in finding solutions to problems. Each "PROBLEM SOLVER" is the first of its kind ever produced in its field. It is the product of a massive effort to illustrate almost any imaginable problem in exceptional depth, detail, and clarity. Each problem is worked out in detail with a step-by-step solution, and the problems are arranged in order of complexity from elementary to advanced. Each book is fully indexed for locating problems rapidly.

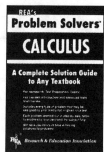

ACCOUNTING
ADVANCED CALCULUS
ALGEBRA & TRIGONOMETRY
AUTOMATIC CONTROL
   SYSTEMS/ROBOTICS
BIOLOGY
BUSINESS, ACCOUNTING, & FINANCE
CALCULUS
CHEMISTRY
COMPLEX VARIABLES
DIFFERENTIAL EQUATIONS
ECONOMICS
ELECTRICAL MACHINES
ELECTRIC CIRCUITS
ELECTROMAGNETICS
ELECTRONIC COMMUNICATIONS
ELECTRONICS
FINITE & DISCRETE MATH
FLUID MECHANICS/DYNAMICS
GENETICS
GEOMETRY
HEAT TRANSFER

LINEAR ALGEBRA
MACHINE DESIGN
MATHEMATICS for ENGINEERS
MECHANICS
NUMERICAL ANALYSIS
OPERATIONS RESEARCH
OPTICS
ORGANIC CHEMISTRY
PHYSICAL CHEMISTRY
PHYSICS
PRE-CALCULUS
PROBABILITY
PSYCHOLOGY
STATISTICS
STRENGTH OF MATERIALS &
   MECHANICS OF SOLIDS
TECHNICAL DESIGN GRAPHICS
THERMODYNAMICS
TOPOLOGY
TRANSPORT PHENOMENA
VECTOR ANALYSIS

*If you would like more information about any of these books,
complete the coupon below and return it to us or visit your local bookstore.*

RESEARCH & EDUCATION ASSOCIATION
61 Ethel Road W. • Piscataway, New Jersey 08854
Phone: (732) 819-8880   **website: www.rea.com**

Please send me more information about your Problem Solver books

Name _____

Address _____

City _____ State _____ Zip _____